TUBE BENDING SIMPLIFIED

TODD WEILER

ISBN: 979-8-89419-958-0 (sc)
ISBN: 979-8-89419-959-7 (hc)
ISBN: 979-8-89419-960-3 (e)

THE EWINGS PUBLISHING

One Galleria Blvd., Suite 1900, Metairie, LA 70001
(504) 702-6708

	1/32	.03125		17/32	.53125
1/16		.0625	9/16		.5625
	3/32	.09375		19/32	.59375
1/8		.125	5/8		.625
	5/32	.15625		21/32	.65625
3/16		.1875	11/16		.6875
	7/32	.21875		23/32	.71875
1/4		.25	3/4		.75
	9/32	.28125		25/36	.78125
5/16		.3125	13/16		.8125
	11/32	.34375		27/32	.84375
3/8		.375	7/8		.875
	13/32	.40625		29/32	.90625
7/16		.4375	15/16		.9375
	15/32	.46875		31/32	.96875
1/2		.5	1		1

Table 1 - Decimal Equivalents

Through the use of any computer, a YouTube video can be viewed which will provide a glimpse into the easiest way to do any bending. After starting any computer, delete the information in the address bar, and in the blank space type in ***www.youtube.com/t04041951*** and then press enter. The video which will play will make the information contained in this book much easier to understand!

TABLE OF CONTENTS

PREFACE

This book has been designed and laid out to be as compact, as clear, and as simple as possible. It will take most of the mystique out of precision tube bending. By using this book as a reference everyone will be able to master the art of precision tube bending. Every bend will look professionally bent.

In any endeavor, knowledge is the single most important key to success. My primary goal in writing this book was to help the mechanic gain knowledge and confidence to accurately bend tube. This knowledge will help you to avoid even the simplest of mistakes thus avoiding cost over-runs for all concerned.

Although it is basic human nature to not find the time to better oneself by keeping up with the best information possible about our jobs until our jobs are threatened; it is a wise person who will have the desire and make the necessary time to constantly teach oneself to learn new skills to stay ahead of progress so that their future and their children's future will always be secure.

In this book, the word tube or any variation of tube; means tube, conduit or pipe. Being able to properly bend tube is a tool in your toolbox of knowledge to help you to get a job or to keep a job. In order to do the best job possible while bending tube, at their fingertips, the tube mechanic needs the best information he or she can get; that information will only be found in this book!

Are you an auto mechanic, a boilermaker, a car frame builder, an electrician, a pipefitter, a plumber, a refrigeration mechanic, a steamfitter, a welder, or simply a person who is trying to expertly bend any tube, conduit, or pipe for any reason? This easy to understand book will benefit everyone who would like to quickly and easily complete any bending project with precision!

This information has been accumulated over a period of more than thirty years of working with tube. Principles and techniques illustrated in this book are ideas which others have taught to me; I in turn have researched, added to, and taught this information to a number of other people. Once you become a good tube bender, you will have all of the basic knowledge required of a great pipefitter mechanic. Keep in mind that what goes around comes around, so always be willing to share knowledge acquired from others with others.

Through the use of this information, each person will learn how to do it right the first time! As soon as measurements have been taken and an isometric plan is sketched, you will be able to bend your sketch without having to resort to any confusing math or time consuming layout methods. If you are going to make any series of bends which initially appear to be puzzling, to clarify the layout in your mind, always make a rough sketch before doing any calculations or making any bends.

By using these simplified concepts, not only will you become a good tube bender but also a great tube layout mechanic! While the old method of measure, mark, and bend does work; this book will explain an easier, more exacting method of tube bending. Simply put, all marks for all bends will be placed onto the tube; then the tube will be bent!

Prior to bending the tube you will place 5 marks on the tube at each bend. After the entire tube is marked, only 3 of these marks, the 2 outside marks and the mark dead center between the 2 outside marks will completely encircle the tube at each bend. A tube ferrule could be used as a mini wrap-around guide to accurately encircle the tube. Three lines are used at each bend so that you will know to place the "0" mark of the bender on one outside line, bend over a centerline, and stop the bend at the 3rd line. When you are orienting the tube, before each bend; always use a bending template to help place the bender into the proper plane prior to the actual bend. After any bend, as long as you can see 3 lines at the next bend; the next bend in succession that needs to be bent will always be obvious and understood. Any bend which is marked in this fashion can be bent as either a left-hand bend or a right-hand bend and the centerline encircled mark on each bend will occur over the visual center of each bend.

Using this method, regardless of whether 1 tube or 200 tubes are measured, marked, and bent; all measurements for all bends will be identical! For field tube bending, this method is far superior to any other tube bending method.

Everyone who owns a company is concerned with three major items, making money and saving money; while at the same time creating a safe, hazard free work environment for their employees. For employers, bending tube is cost effective; while involving limited safety hazards for employees!

More employers are beginning to realize the following benefits of tube bending, which are only a few of the many reasons that bends will replace some welds on various jobs. A weld joint which can crack and leak, takes considerably longer to prepare and weld, than a similar bend takes to layout and bend; not even taking into account for probable weld repair time. Weld repair by itself, makes a bend more economical and more practical to fabricate and install whenever feasible. In contrast to a weld joint which can be weakened due to too much heat from too much welding; a bend is almost as strong as the base metal which is in the rest of the tube. No pre-heating or purging, both probable safety hazards which will need to be mitigated; is required with any bend. Every bend can be made to varying degrees, while unless special ordered, weld fittings are standard 90 degree or 45 degree fittings. Any standard weld fitting can be mitered to any degree, this mitering increases weld preparation time considerably! Also; less turbulence is transmitted to the product in the tube when it is bent rather than welded, resulting in a decreased resistance to flow with a resulting probable need for less expensive transmission pumps.

With the tube bending tools which are presently on the market; the vast majority of tube, up to 2" schedule 40 pipe size, can easily be field fabricated and installed saving money and time for both the employer and the customer. The job of the tube bender is first to be safe, and second to do the best job possible while always ensuring that the employer's and the customer's bottom line costs are always foremost in mind.

Worldwide; the demand for skilled tube benders is constantly increasing. Everyone who knows how to skillfully bend tube, and who has the proper tube bending tools can easily bend any size tube in or for any application.

BEGINNING TUBE MARKING

This book contains many useful and valuable points of reference. If you are already an accomplished tube bender, this knowledge and these techniques may be a useful reference to you. If you are new to bending and you want to learn the best way to do any bending, this book should become your bending bible!

It has been said that a picture is worth a thousand words. In keeping with that slogan, each section in this book will begin with an illustration, followed by an explanation of that illustration.

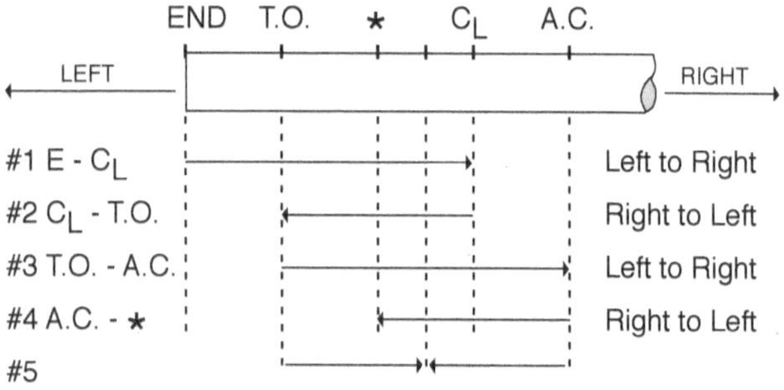

Figure 1 - 5 Mark Layout

Getting started is the hardest part of any journey. With tube bending our journey will begin with an explanation of tube marking, followed by an overview of this method of tube bending.

For demonstration clarity, most explanations and illustrations in this book will pertain to 3/8" outside diameter (O.D.) tube, a tube bender with a 15/16" bend radius, and most bends will be described as 90 degree bends. The information contained herein would be used as a guide from which all other size tube layouts would be computed.

The two formulas required to position all marks onto the tube after the initial centerline mark for any degree of bend are:

1. Take-off (T.O.) = the tangent of ½ of the degrees of the bend x the bender radius.

2. Arc Length (A.C.) = the degree of the bend x 0.01745 x the bender radius.

The Arc length equals the total length of any bend and 0.01745 is the tangent of 1 degree.

The preceding T.O. formula will only work up through a 90 degree bend. **Never use the preceding T.O. tangent formula for any degree of bend that is greater than 90 degrees.** For any bend that is greater than 90 degrees the T.O. measurement is accumulative (beginning with the measurement for one 90 degree bend)! Two examples for the T.O. measurement follow: Example #1, the T.O. measurement of a 180 degree bend equals the T.O. measurement of (2) 90 degree bends. Example #2, the T.O. measurement of a 108 degree bend equals the T.O. measurement of (1) 90 degree bend plus the T.O. measurement of (1) 18 degree bend.

Since the T.O, measurement is accumulative from 91 degrees to 180 degrees; in the same manner, the A.C. and Tube Gain (T.G. – Figure 2 page 15) measurements are also accumulative from 91 degrees to 180 degrees.

The length of the measurement on the tube between the reference mark and the A.C. mark will always be the same as the length of the measurement on the tube between the T.O. mark and the centerline mark. After the centerline, T.O., A.C., and reference marks are marked onto the tube; the reference mark is the point on the tube to start the next center to center measurement, or center to end of tube measurement. Before moving on to the next bend in succession, you could bisect the distance on the tube from the T.O. mark to the A.C. mark to arrive at mark number 5; the visual center of each bend, or you could wait and place Mark #5 onto the tube while you are doing tube encircle marking.

The easiest way to get the correct measurements to mark the tube before bending it is to look at the tube bender to determine its' tube bending radius. Then locate the proper chart in this book to get the correct T.O. and A.C. measurements for the bend that you will be making. Finally, make yourself a tubespoon marker so that all bends will be accurately and identically marked.

No book could contain a chart or an illustration for every conceivable bend imaginable with every bend radius, that is why the two formulas along with the charts and illustrations in this book are so useful; they are templates for you to follow so that you can determine conclusively any measurement in any configuration which will be needed for any bend!

BEND RADIUS CHARTS

It is more important to know the radius of the bender which will be used to make any bend than it is to know the diameter of the tube which is being bent! When using any tube bender make certain that the tube fits snugly into the tube bender groove, not too tight and not too loose. Too tight a fit and the tube will not fit into the bender, making the bend impossible to make; while too loose a fit could flatten, kink, or wrinkle a bend of only a few degrees. A tube which is bent using the proper sized tube bender will not have any kinks or wrinkles throughout the radius of the bend and all areas throughout the bend will contain adequate tube material to contain the pressure within the tube.

In the charts in this section, the decimal representation of the multiplied answer is rounded from 100th of an inch to the nearest 1/16" (Table 1). For any degree of bend, the T.O. measurement x 2 will always be greater than or equal to the A.C. measurement. Due to conversion from 100th of an inch to a 16th of an inch, some numbers will require minor rounding down. Rarely round down the T.O. mark number, if one of the two numbers needs to be rounded down; round down the A.C. mark number. All numbers which are rounded in the following charts will be designated by highlighting the number. The following charts may be used for any size tube which corresponds to that designated tube bend radius; they illustrate the 7 most commonly used degrees of bends. All other degree of bends, and all bend radiuses not shown, may be calculated using the two formulas on page 7 along with tangents from the Trigonometry Tables. After acquiring the needed T.O. and A.C. measurements, a tubespoon marker should be constructed to transfer accurate marks onto all tubes.

	3/8"		7/16"		9/16"		5/8"		11/16"		3/4"	
of Bend	T.O.	A.C.	T.O.	A.C.	T.O.	A.C.	T.O.	A.C.	T.O.	A.C.	T.O.	A.C.
90	3/8	9/16	7/16	11/16	9/16	7/8	5/8	1	11/16	1 1/16	3/4	1 3/16
75	5/16	1/2	5/16	9/16	7/16	3/4	1/2	13/16	1/2	7/8	9/16	1
60	3/16	3/8	1/4	7/16	5/16	9/16	3/8	5/8	3/8	3/4	7/16	13/16
45	**1/8**	**1/4**	3/16	5/16	1/4	7/16	1/4	1/2	5/16	9/16	5/16	9/16
30	1/8	3/16	1/8	1/4	1/8	**1/4**	3/16	5/16	3/16	3/8	3/16	3/8
22.5	1/16	1/8	1/16	**1/8**	1/8	1/4	1/8	1/4	1/8	1/4	1/8	**1/4**
15	1/16	1/8	1/16	1/8	1/16	1/8	1/16	**1/8**	1/16	**1/8**	1/8	3/16

Chart 1 - Bend Radius

	15/16"		1"		1 1/8"		1 1/4"		1 1/2"		1 3/4"	
of Bend	T.O.	A.C.	T.O.	A.C.	T.O.	A.C.	T.O.	A.C.	T.O.	A.C.	T.O.	A.C.
90	15/16	1 1/2	1	1 9/16	1 1/8	1 3/4	1 1/4	1 15/16	1 1/2	2 3/8	1 3/4	2 3/4
75	3/4	1 1/4	3/4	1 5/16	7/8	1 1/2	15/16	1 5/8	1 1/8	1 15/16	1 5/16	2 5/16
60	9/16	1	9/16	1 1/16	5/8	1 3/16	3/4	1 5/16	7/8	1 9/16	1	1 13/16
45	3/8	3/4	7/16	13/16	7/16	7/8	1/2	1	5/8	1 3/16	3/4	1 3/8
30	1/4	1/2	1/4	1/2	5/16	9/16	5/16	5/8	3/8	**3/4**	1/2	15/16
22.5	3/16	3/8	3/16	3/8	1/4	7/16	1/4	1/2	5/16	9/16	3/8	11/16
15	1/8	1/4	1/8	1/4	1/8	**1/4**	3/16	5/16	3/16	3/8	1/4	7/16

Chart 2 - Bend Radius

	1 7/8"		2"		2 1/4"		2 1/2"		2 5/8"		3"	
of Bend	T.O.	A.C.	T.O.	A.C.	T.O.	A.C.	T.O.	A.C.	T.O.	A.C.	T.O.	A.C.
90	1 7/8	2 15/16	2	3 1/8	2 1/4	3 9/16	2 1/2	3 15/16	2 5/8	4 1/8	3	4 11/16
75	1 7/16	2 7/16	1 9/16	2 5/8	1 3/4	2 15/16	1 15/16	3 1/4	2	3 7/16	2 5/16	3 15/16
60	1 1/16	1 15/16	1 1/8	2 1/8	1 5/16	2 3/8	1 7/16	2 5/8	1 1/2	2 3/4	1 3/4	3 1/8
45	3/4	1 1/2	13/16	1 9/16	15/16	1 3/4	1 1/16	1 15/16	1 1/16	2 1/16	1 1/4	2 3/8
30	1/2	1	9/16	1 1/16	5/8	1 3/16	11/16	1 5/16	11/16	1 3/8	13/16	1 9/16
22.5	3/8	3/4	3/8	**3/4**	7/16	7/8	1/2	1	1/2	1	5/8	1 3/16
15	1/4	1/2	1/4	1/2	5/16	9/16	5/16	5/8	3/8	11/16	3/8	**3/4**

Chart 3 - Bend Radius

	3 1/4"		3 3/8"		3 1/2"		3 3/4"		4"		4 1/2"	
of Bend	T.O.	A.C.	T.O.	A.C.	T.O.	A.C.	T.O.	A.C.	T.O.	A.C.	T.O.	A.C.
90	3 1/4	5 1/8	3 3/8	5 5/16	3 1/2	5 1/2	3 3/4	5 7/8	4	6 5/16	4 1/2	7 1/16
75	2 1/2	4 1/4	2 9/16	4 7/16	2 11/16	4 9/16	2 7/8	4 15/16	3 1/16	5 1/4	3 7/16	5 7/8
60	1 7/8	3 3/8	1 15/16	3 9/16	2	3 11/16	2 3/16	3 15/16	2 5/16	4 3/16	2 5/8	4 11/16
45	1 3/8	2 9/16	1 3/8	2 5/8	1 7/16	2 3/4	1 9/16	2 15/16	1 11/16	3 1/8	1 7/8	3 9/16
30	7/8	1 11/16	7/8	1 3/4	15/16	1 13/16	1	1 15/16	1 1/16	2 1/8	1 3/16	2 3/8
22.5	5/8	1 1/4	11/16	1 5/16	11/16	1 3/8	3/4	1 1/2	13/16	1 9/16	7/8	1 3/4
15	7/16	7/8	7/16	7/8	7/16	15/16	1/2	1	1/2	1	9/16	1 1/8

Chart 4 - Bend Radius

	5"		6"		8"		11"	
° of Bend	T.O.	A.C.	T.O.	A.C.	T.O.	A.C.	T.O.	A.C.
90	5	7 7/8	6	9 7/16	8	12 9/16	11	17 1/4
75	3 13/16	6 9/16	4 5/8	7 7/8	6 1/8	10 1/2	8 7/16	14 3/8
60	2 7/8	5 1/4	3 7/16	6 5/16	4 5/8	8 3/8	6 3/8	11 1/2
45	2 1/16	3 15/16	2 1/2	4 11/16	3 5/16	6 5/16	4 9/16	8 5/8
30	1 5/16	2 5/8	1 5/8	3 1/8	2 1/8	4 3/16	2 15/16	5 3/4
22.5	1	1 15/16	1 5/16	2 3/8	1 9/16	3 1/8	2 3/16	4 5/16
15	11/16	1 5/16	13/16	1 9/16	1 1/16	2 1/8	1 7/16	2 7/8

Chart 5 - Bend Radius

Footnotes:

EMT is Electrical Metallic Tubing

1/2" EMT has a 5" Radius

3/4" EMT has a 6" Radius

1" EMT has an 8" Radius

1-1/4" EMT has an 11" Radius

TUBESPOON MARKER

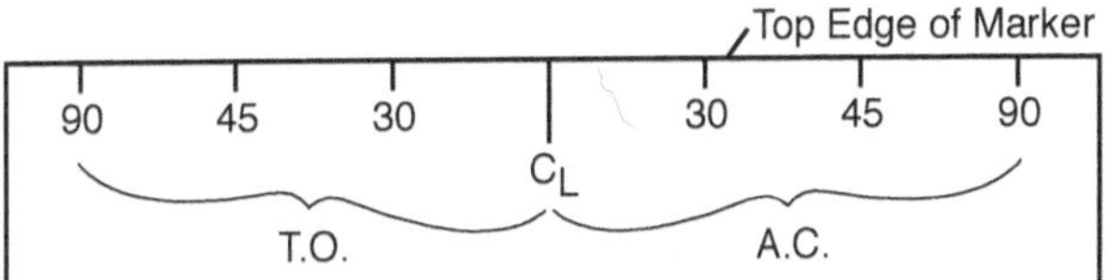

	T.O.	A.C.	T.G.
90	15/16	1 1/2	3/8
45	3/8	3/4	0
30	1/4	1/2	0

Figure 2 - Tubespoon Marker
Not to scale

To calculate tube gain for any bend with any bend radius, T.G. = (T.O. x 2) – A.C.

Always make and then use a tubespoon marker to mark each bend on the tube prior to doing any bending. The tubespoon marker is a simple yet helpful tool to assure that all marked measurements are identical. To make a tube-spoon marker use a blank index card or anything else which you can write on. The focal point of the tubespoon marker will be the centerline. The marks for all bends for the T.O. will be to the left of the centerline and the A.C. marks will be to the right of the centerline. The most often used degree of bends will be located in the tube bending charts in this book or you may calculate your own degree marks using the two formulas on page 7.

As you are facing the tube, always measure and start marking the tube from the left end of the tube. Once the first end to centerline is marked onto the tube, the centerline mark on the tubespoon marker will be laid over the centerline mark on the tube. The marks for any degree of bend will be transferred from the tubespoon marker onto the tube. After the T.O. and A.C. marks for any degree of bend are marked onto the tube, move the tubespoon marker slightly to the right until the centerline mark on the tubespoon marker is over the A.C. mark which was just marked onto the tube. The T.O. mark on the tubespoon marker for that degree of bend is the place to make a mark onto the tube for the reference mark (*), Mark #4. This is the starting centerline mark for the next measurement in succession on the tube. The reference mark takes into account for tube gain at each bend regardless of whether any bend will be bent as a left-hand bend or a right-hand bend. Before moving on to the next bend in succession, you could bisect the distance on the tube from the T.O. mark to the A.C. mark to arrive at mark number 5; the visual center of each bend, or you could wait and place Mark #5 onto the tube while you are doing tube encircle marking.

Use of the tubespoon marker will ensure that each bend on each tube is marked identically. This works much better than measuring each individual measurement each time with a rule. Since there is less chance of reading the tubespoon marker incorrectly, there will be less chance of producing a marking error.

The most important measurement to put on any tube-spoon marker is the T.O. measurement for a 90 degree bend. At a glance, that measurement will reveal to anyone the radius of the tube bender that this tubespoon marker was made for.

After the tube is bent, from either the T.O. mark line or the A.C. mark line to the center of the intersecting tube run will equal the T.O. measurement for that degree of bend.

If you need to measure from an existing 90 degree bend and you do not know the radius of the bender which was used while making that existing 90 degree bend, put a straight edge on the back side of the tube before the bend and subtract ½ of the diameter of the tube and then measure to the centerline of the next tube bend. (Refer to Figure 5 page 26 to see the centerline of a tube run.) Then use your tubespoon marker and proceed as though you were marking any other tube to bend.

BASIC EXPLANATION OF THE FIVE MARK METHOD

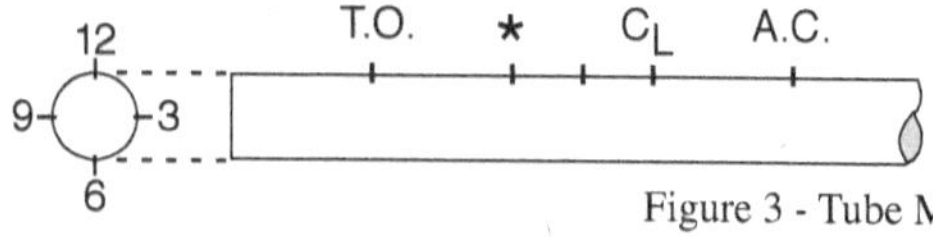

Figure 3 - Tube Marking

Just as a child must walk before learning to run, a good tube bender must first learn the best way to mark the tube and then put the marks onto the tube in the proper sequence prior to bending. Once you have a series of measurements for a tube layout, precise measuring and marking become critical to the tube bending process. Carry out the following 10 steps in progression to ensure the greatest marking accuracy.

1. Lay the tube to be marked horizontally on a table in front of you. For most people a comfortable table height would be 36".

2. As you are facing the tube, look at the left open end of the tube as though it were the face of a clock.

3. Use 12 o'clock (the very top edge) as an imaginary line along the entire outside length of the tube to place all marks for all bends in succession onto the tube.

4. With measurements for your first tube to be bent in hand, start at the left end of the tube and mark your first end to centerline measurement onto the tube.

5. After marking the first centerline measurement onto the tube, position the centerline mark of the tubespoon marker over the centerline mark on the tube, then transfer Mark #2 (T.O.) and Mark #3 (A.C.), from the tubespoon marker onto the tube.

6. Immediately slide the tubespoon marker slightly to the right until the centerline mark on the tubespoon marker lines up with the A.C. mark which you have marked onto the tube. The T.O. measurement on the tube-spoon marker is the point on the tube to place the reference mark (*), Mark #4.

7. Use the reference mark on the tube as the starting point for the next measurement in succession.

8. Place the next set of 5 marks onto the tube.

9. Keep proceeding in this fashion until the entire tube is marked.

10. For the last mark on the tube at each bend, bisect Mark #1 (the centerline mark) and Mark #4 (the reference mark - *) with a mark. This will be the 5th mark on the tube and will designate the visual center of the bend after the bend is bent regardless of whether a left-hand bend or a right-hand bend is bent. This point is dead center between the T.O. mark and the A.C. mark.

OVERVIEW OF THE FIVE MARKS

Mark #1 is the centerline mark measurement. For initial layout purposes, the first centerline measurement on the tube (from the left end of the tube) is an end of tube to centerline measurement.

Mark #2 is the T.O. mark measurement and is used when making all left- hand bends. The left-hand bend is always the bend of preference. The formula required to arrive at the T.O. mark is:

T.O. = the tangent of ½ of the degrees of the bend x the bender radius.

Example: T.O. mark of a 90 degree bend using a 15/16" radius bender.

T.O. = the tangent of 45 degrees x 15/16

T.O. = 1 (from the Trigonometry Tables) x 0.9375 (from Table 1)

T.O. = 0.9375 (15/16 – Table 1)

This measurement, centerline to T.O., originates at the centerline and measures from the right to the left on the tube. All left-hand bends will start to bend at the T.O.

mark and will proceed to bend over the centerline mark and will stop at the A.C. mark. Prior to a left-hand bend, 3 points will align precisely. The T.O. mark which was placed onto the tube will be aligned with the "0" mark on the radius block of the bender. Then the "0" mark on the bender handle will be aligned with the T.O. mark on the tube and the opposing "0" mark on the bender radius block. The latch on the tube bender will then be flipped into place to lock the tube into the bender.

The measurement from the centerline of the intersecting tube run to the T.O. for any tube at a 90 degree bend will equal the radius of the tube bender used with the tube. This measurement is generally stamped into the surface of all hand benders or it may be measured from the pivot center point of the tube bender to the centerline of the tube groove on the bender.

Mark #3 is the A.C. mark measurement and is used when making all right-hand bends. This measurement originates at the T.O. mark and measures from the left to the right over the centerline mark to arrive at the A.C. mark. There will be instances when it will not be possible to make all bends (marked on the tube) with the tube positioned from the same end relative to the bender; it will then become necessary to make a right-hand bend.

One of two things must be done. One, the tube will be turned end for end in the bender, (this practice must be done with a table mounted bender) or two, the bender will be flipped end for end (180 degrees) on the tube. This method will be used with a portable hand bender. The bend will start at the A.C. mark, bending over the centerline mark, and the bend will stop at the T.O. mark.

All right-hand bends, also known as reverse or back bends, will originate at the A.C. mark, will proceed to bend over the centerline encircled mark, and will stop at the T.O. mark. Prior to a right-hand bend, 3 points must align precisely. The A.C. mark which was placed onto the tube will be aligned with the "0" mark on the radius block of the bender; The "0" mark on the bender handle will be aligned with the A.C. mark on the tube and the opposing "0" mark on the radius block, and then the latch will be locked into place. The following is an example of how to use math to accurately arrive at the precise A.C. point on the tube, this formula will work with any size tube bender and with any degree of bend.

Example: A.C. mark of a 90 degree bend using a bender with a 15/16" radius.

A.C. = the degrees of the bend x .01745 x the bender radius.

A.C. = 90 x .01745 x 15/16

A.C. = 90 x .01745 x .9375

A.C. = 1.4723 (Table 1 [1 ½"])

Mark #4 is the reference mark (*), this is the beginning centerline mark for the next measurement. This mark is mainly used as a starting centerline reference for the next measurement to establish the next set of 5 marks on the tube or to establish the end of the tube. This reference centerline mark originates at the A.C. mark and is measured from the right to the left on the tube (A.C. to *).

Measurement #4 (A.C. mark to reference mark - *) is always the same measurement as measurement #2, (centerline mark to T.O. mark). This mark on the tube will be the start of a centerline to centerline measurement for the next measurement in succession or for the centerline to the end of the tube measurement.

Mark #4 (*) takes into account for tube gain for the degree of the bend which is being marked; because this reference mark takes into account for tube gain, it does not make a difference whether any bend is bent as a left-hand bend or a right-hand bend.

Mark #5 will be the visual center of the bend after the tube is bent regardless of whether the bend was bent as a left-hand bend or a right-hand bend. To reduce confusion, due to the numerous marks which should be placed onto the tube at each bend to maintain precision tube marking; this mark should not be marked onto the tube until tube encircle marking is being done.

All bends will originate with the "0" mark of the bender radius block on either outside encircled mark of the bend on the tube, bending over the centerline encircled mark and stopping at the third encircled mark.

For some of the smaller radius benders with the smaller degree of bends, and for any 180 degree bend; marks #1, 4, and 5 will be at the same point on the tube. All five marks are equally important for a precise tube layout. After the tube is bent, from either outside mark on the tube (the T.O. mark or the A.C. mark) at any bend to the exact center of the intersecting tube run will equal the T.O. dimension of the degree of that bend.

TUBE ENCIRCLE MARKING

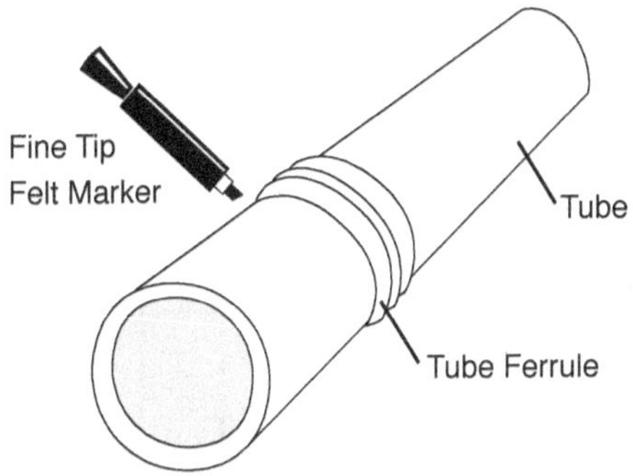

Figure 4 - Tube Encircle Marking

After the entire tube is marked, 3 marks at each bend will be remarked to completely encircle the tube. Use a fine tip magic marker along with a tube ferrule as a mini wrap-around guide and completely encircle the tube at the T.O. mark, the A.C. mark, and the point dead center between these two marks.

Mark #5, the visual center of any bend should not be marked onto the tube until tube encircle marking is being done. This is the center mark of the 3 encircled marks at each bend, since it is a visual mark, it could be placed by eyeing its' position on the tube.

TUBE GAIN

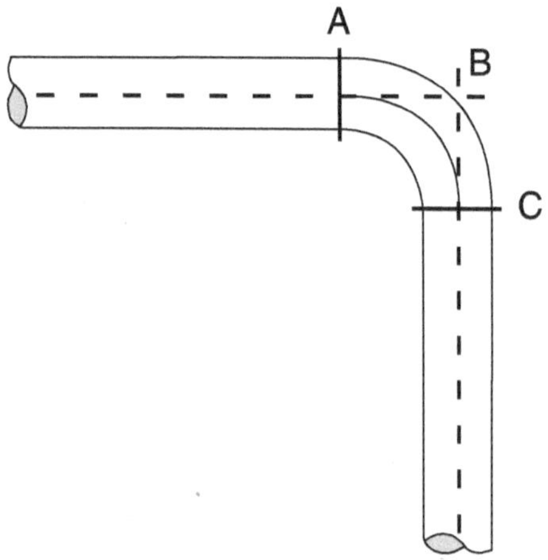

Figure 5 - Tube Gain

AB = the end of the bend to the centerline.

BC = the centerline to the end of the bend.

Arc length A.C. = the run of the tube at the centerline of the tube. Due to the fact that Arc length A.C. follows the curve of the tube, it can clearly be seen that Arc length A.C. is a shorter distance than length AB + length BC.

Every time a bend is made in a tube, less end to end tube is needed because the bend takes a curved shortcut across the inside angle of the intersection of the two tube runs.

The best way to determine tube gain for any bend is to use the following method.

Tube gain = (T.O. x 2) – A.C.

Example: A 90 degree bend using a bender with a 15/16" radius.

Tube gain = (15/16" x 2) – 1 ½"

Tube gain = (1 - 7/8") -1 ½"

Tube gain = 3/8"

For any degree of bend with any radius bender T.O. x 2 will always be greater than or equal to A.C.

By using this tube gain method along with a tube layout schedule, the end to end length of any tube needed for any tube run may be determined to within 1/16" prior to bending.

Each of the five marks on the tube originates from the preceding mark in a logical progression.

BENDING TIPS

1. When bending, always try to bend with the same end of the tube relative to the bender. It will be less confusing! Since the tube was measured and marked from the left, always try to make left-hand bends.

2. Planning, layout, and final measuring are time consuming and critical. It is important to not become distracted. A momentary lapse in concentration can prove costly both in time and in material. Keep your concentration on the work at hand.

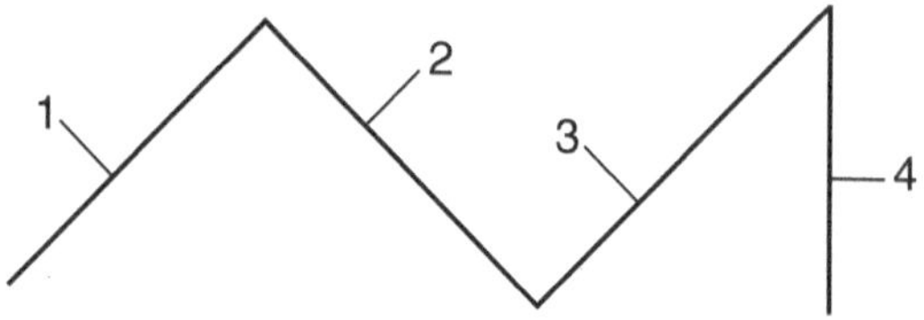

Figure 6 - Isometric Tube Layout

3. After planning the work, laying out the tube route, and getting the necessary measurements together; make a rough isometric drawing of the tube layout. This isometric drawing does not need to be to scale. In the preceding drawing, all bends are illustrated at 90 degrees simply for clarity.

BENDING TEMPLATE

Always keep in mind the importance of tube bending direction. Prior to the actual bending of the tube, use the isometric drawing (Figure 6) and make a bending template. Use a 1/16" welding filler rod, or something similar, cut approximately 12" long. The template, although not to scale, will ensure proper visualization and the correct perspective of each bend in order.

It is far easier and less costly to rebend another welding rod than it is to rebend another tube. This simple practice will guarantee less tube waste, because the tube will be properly oriented prior to each bend.

TUBE LAYOUT SCHEDULE

	#1 E-C	#2 C-C	#3 C-C	#4 C-E
Tube From Figure 6	2 1/2" - 90	3" - 90	8" - 90	6"
Next Tube	?	?	?	?
Next Tube	?	?	?	?

Table 2 - Tube Layout Schedule

In the tube layout schedule above, each section includes the length of that section along with the degree of the bend of that section of the tube.

A tube layout schedule is important, it reduces the amount of measuring time when laying out multiple tubes, either together as in a bank of tubes, or in a tray where the tubes are bound tightly together.

As soon as the measurements for the first tube have been physically taken; all measurements for tubes in succession, in the schedule, would be calculated due to their relative position from the first tube. All that needs to be done is to add or to subtract from the center of the first tube to the center of the tube that measurements are needed for. Measurements for all tubes in the schedule would be calculated and inserted into the schedule using this method.

Final appearance will become readily apparent; a tube mechanic with the proper knowledge and tools will accomplish a neat, orderly tube layout. The end result being a job well done and pride in one's work. Upon satisfactorily learning this method, a good tube bender will be a major asset on any job and will never use the method of measure, bend, measure, bend, to bend tube again. This five mark method achieves the best of precision tube bending.

DOUBLE CHECKING

Prior to each bend, use the tube layout schedule to recheck all measurements to ensure that no bend marks were missed and that all bends were marked with the proper degree line marks. Next, use a tube ferrule and a fine tip felt marker to encircle the tube at the T.O., center of bend, and A.C. marks at each bend. Use a narrow line marker because too wide a mark will compromise good measurements.

This is the time to make the first bend. Always check with the schedule to verify each bend measurement and the degree of the bend. Use a bending template to visually rotate the tube into the proper bend plane. Now, make your first bend! Proceed in this fashion; checking, rotating, and bending until all of the bends are made in the tube.

This practice of double checking may appear to be overly meticulous, time consuming, and somewhat ridiculous; however, it will alleviate confusion, general mistakes, and missed bends. One minor mistake or wrong bend will add considerable time and cost to a project. Once this method is thoroughly learned, it will prove to save time and assure maximum accuracy for every bend. Proper tube bending takes extreme concentration, so take your time and do an excellent job.

After the tube is bent but before moving onto the next tube in the schedule, draw a pencil line through all of the measurements on the schedule for that tube. This practice will prove beneficial when there are two or more tubes listed in the schedule. It prevents marking the same tube, or a measurement from that tube, a second time.

As soon as the entire tube is bent, slip a tube union onto each end of the tube until the tube bottoms onto the shoulder of the tube union. Using a fine line magic marker, make a mark completely around the tube. In the field when final tube assembly is being completed, if this line is only slightly visible, the person doing final tube fitting tightening will be assured that the tube is properly bottomed into the fitting. This will help to guarantee a leak free connection.

If you are positive that the tube will fit when you carry it from the bench to where it will be installed; on the bench, the tube ferrule could be preset with the nut. Final field assembly will be quicker and easier!

TUBE FITTING MAKE-UP

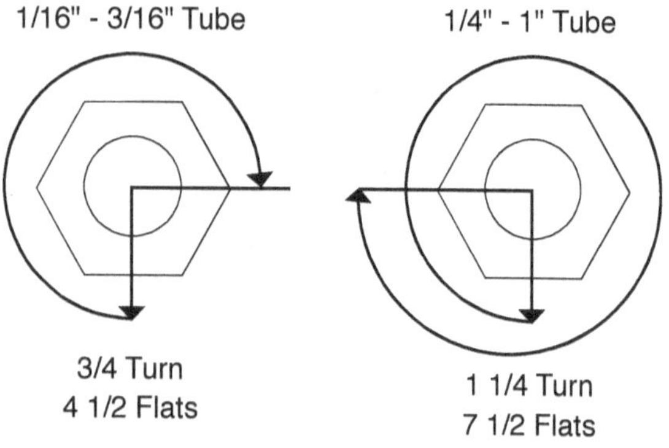

Figure 7 - Tube Fitting Make-up

Tube fitting gauges (go-no-go gauges), while being precision tools, have been known to give a false sense of proper tube fitting tightening. Because of considerable variations in tube wall thickness and surface hardness, instrument tube fittings are tightened due to compression, not due to torque. If a compression fitting is tightened too tightly; not only will it be next to impossible to take the tube fitting apart at a later date, it may also make the fitting more prone to leakage. Follow this simple four step instruction to tighten instrument fittings.

1. After using a clean rag to clean the place on the tube where the fitting will be tightened, insert the tube into the fitting.

2. Make certain that the tube rests firmly onto the shoulder of the tube fitting body and that the nut is finger tight.

3. Mark the nut at any convenient spot so that there is a visual reference.

4. While holding the fitting body steady with one wrench, tighten the nut with another wrench the proper number of flats.

TANGENT OF ANGLE

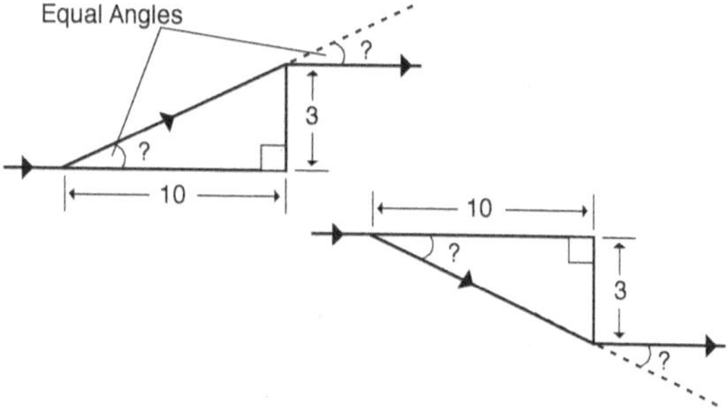

Figure 8 - Straight Line Offset

All tube offsets are based on right triangles. Tangent of angle is the formula needed to determine any offset angle of bend. Tangent of angle is determined by dividing the set by the run. In a right triangle, the set is the side opposite the reference angle. The run is the shortest side adjacent to the reference angle. The angle of the bend is the number of degrees from a straight line that the tube changes direction. To form an offset bend it is necessary to make two back to back equal bends while proceeding in one direction.

Any angle may be found for any unknown bend by using the following tangent of angle formula along with tangents from the Trigonometry Tables.

Example: What is the degree of the unknown angle illustrated in figure 8? In figure 8 the tube is traveling horizontally, where it then bends up or down. The set measures 3 and the run measures 10.

Tangent of angle = set/run (opposite side/adjacent side)

Tangent of angle = 3/10

Tangent of angle = 0.3

Now using the tangents from the Trigonometry Tables, the nearest whole degree to 0.3 is 17 degrees. This is the angle of each of the bends of the offset after which the tube will continue parallel to its' initial run. Any angle for any degree of bend will be calculated in this fashion.

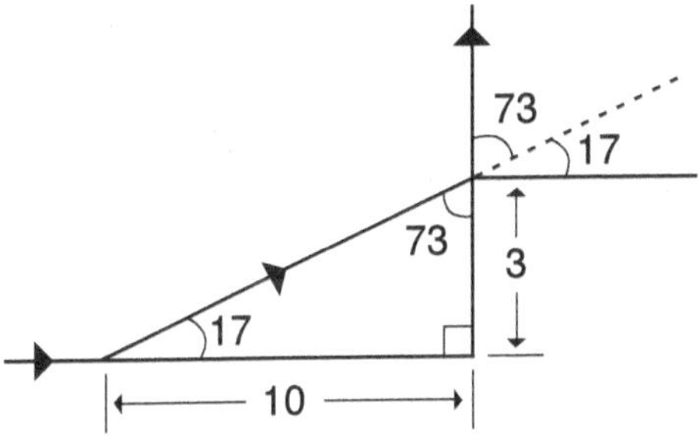

Figure 9 - 90° Offset

In figure 9, if after the second bend is made, the tube proceeds at a 90 degree plane from the initial direction of travel, we can easily calculate the degree of the second bend. There are 180 degrees of bend in every triangle, minus one 90 degree right angle, leaves 90 degrees of bend for the remaining two bends. Our first angle had 17 degrees of bend, if we subtract that from the remaining 90 degrees in a right triangle, it is easy to calculate that our second bend is 73 degrees.

The tangent formula utilized during tube layout, along with the decimal representation from the tangents in the Trigonometry Tables, should be used to copy an existing layout when that practice is necessary.

METRIC TUBE BENDING

This book has been formatted using the SAE (non-metric) measuring system. To bend metric tube which is represented in millimeters, a metric tube bender is used and its' radius is expressed in millimeters.

Any time that it is necessary to convert from inches to millimeters or vice versa, use the following formula:
1 inch = 25.4 millimeters.

The easiest way to convert inches to millimeters is to change the fractional inch measurement into its' decimal equivalent, then multiply that decimal by 25.4 millimeters per inch.

Example: (1 – 5/8") = 1.625 inches multiplied by 25.4 millimeters per inch = 41.275 millimeters.

The easiest way to convert millimeters to inches is to divide the millimeters by 25.4 millimeters per inch and then change that decimal inch measurement into a fractional measurement.

Example: 41.275 millimeters divided by 25.4 millimeters per inch = 1.625 inches (1 – 5/8")

By using the two formulas shown on page 7, it will be easy to construct a metric chart (similar to charts 1-5) for any size metric tube bender.

ADVANCED SKILLS

Now that you understand how easy tube bending is, this section will concentrate on progressing to further tube bending principles. No tube bending manual would be complete without an explanation of tube staggering, tube stacking, tube joining, tube stack setback, and the use of the 4" tray, and ¼" by 1" aluminum flat bar to hold the tube tightly into the tray.

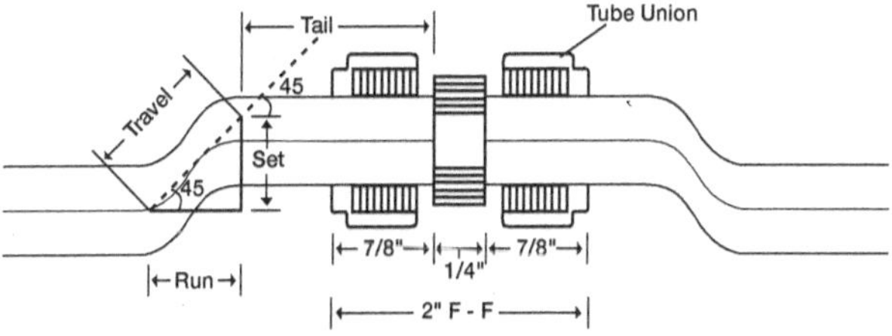

Figure 10 - Tube Union in Offset

In a tube run, the most practical way to join tubes together is with a mechanical tube union. Since tube is manufactured in lengths of 20 feet, offset layouts must be designed into the tube run. These offsets provide adequate space for the installation of tube unions to join tubes together. The preceding example illustrates 3/8" O.D. tube, with two back to back 45 degree offsetting bends, then a tube union, and finally two more back to back 45 degree offsetting bends; with the tube proceeding in the same direction as before any bends were made into the tube run.

It is easy to see that the tube union takes up more space than the actual tube and because it is a mechanical assembly, allocation of space for wrenches to tighten the tube union to the tube must be made. The most eye appealing and the most practical back to back offsetting bends to make would be 45 degrees due to the fact that the set and the run will always be the same length. This will make for a layout which is both easy to design and easy to sketch. Enough circumstances will develop when unusual bends will have to be figured out and calculated into the tube layout. Keep those instances to a minimum. In the tube run, keep offset bends close together to make a finished layout that will be as orderly and as symmetrical as possible. The layout should be uniform on both sides of the tube union(s). Make certain that no tube union is at the same elevation as any straight tube run in the tube run layout.

When installing fittings near tube bends, there must be sufficient straight length of tube to allow the tube to be bottomed into the fitting. The make-up length bottomed into the tube union dictates the minimum length of the tube tail. The tube tail is the end of the tube where the tube union is tightened into place; either after the last 45 degree bend of the first tube or before the first 45 degree bend of the joining tube. Referring to Figure 10 as an example, the length of the tube tail measures make-up, in this case 7/8" for the tube union, plus an additional 1" to equal a 1 – 7/8" tail. The 1" area on the tube, from the nearest 45 degree bend to the tube union will allow a space for an identifying mark to be placed onto the tube to help locate any specific tube, in the tube layout, at any later date.

TRIGONOMETRY

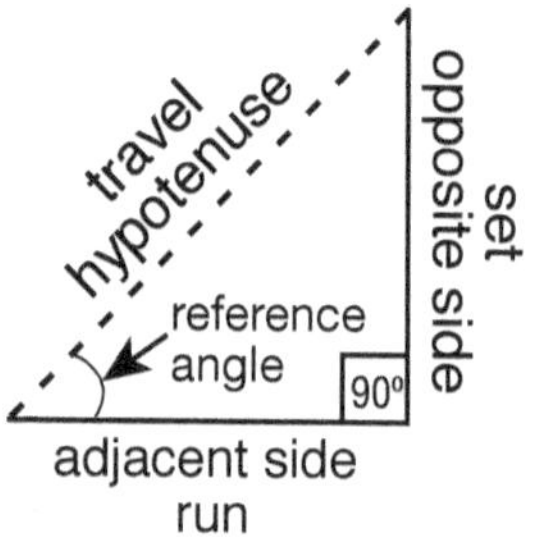

Figure 11 - Right Triangle

Trigonometry deals with solutions to problems involving right triangles. A right triangle is a triangle with one 90 degree corner. All piping offsets are based on the right triangle. The preceding right triangle and the following six trigonometric functions will be used by anyone who is arranging a piping layout.

Set = Run x the Tangent of the angle

Set = Travel x the Sine of the angle

Run = Travel x the Cosine of the angle

Run = Set x the Cotangent of the angle

Travel = Set x the Cosecant of the angle

Travel = Run x the Secant of the angle

Set/Run = Tangent of the angle

Set/Travel = Sine of the angle

Run/Travel = Cosine of the angle

Run/Set = Cotangent of the angle

Travel/Set = Cosecant of the angle

Travel/Run = Secant of the angle

With every triangle which has one 90 degree right angle, as soon as 1 additional angle is known, the remaining angle will equal 90 minus the known angle. After the 3 angles are known; if the length of 1 of the 3 sides is known, use the preceding 6 trigonometric functions, and the following trigonometric tables, along with the smallest angle (the one which is 45 degrees or less), then calculate the length of the other 3 sides. In a right triangle, any angle of the two remaining angles which is greater than 45 degrees, but less than 90 degrees will use a different formula to compute the length of the remaining 2 sides.

Trigonometric Tables

deg	sin(x)	cos(x)	tan(x)	cot(x)	csc(x)	sec(x)
0	0.0000	1.0000	0.0000			1.0000
1	0.0175	0.9998	0.0175	57.2900	57.2987	1.0002
2	0.0349	0.9994	0.0349	28.6363	28.6537	1.0006
3	0.0523	0.9986	0.0524	19.0811	19.1073	1.0014
4	0.0698	0.9976	0.0699	14.3007	14.3356	1.0024
5	0.0872	0.9962	0.0875	11.4301	11.4737	1.0038
6	0.1045	0.9945	0.1051	9.5144	9.5668	1.0055
7	0.1219	0.9925	0.1228	8.1443	8.2055	1.0075
8	0.1392	0.9903	0.1405	7.1154	7.1863	1.0098
9	0.1564	0.9877	0.1584	6.3138	6.3925	1.0125
10	0.1736	0.9848	0.1763	5.6713	5.7588	1.0154
11	0.1908	0.9816	0.1944	5.1446	5.2408	1.0187
12	0.2079	0.9781	0.2126	4.7046	4.8097	1.0223
13	0.2250	0.9744	0.2309	4.3315	4.4454	1.0263
14	0.2419	0.9703	0.2493	4.0108	4.1336	1.0306
15	0.2588	0.9659	0.2679	3.7321	3.8637	1.0353
16	0.2756	0.9613	0.2867	3.4874	3.6280	1.0403
17	0.2924	0.9563	0.3057	3.2709	3.4203	1.0457

Trigonometric Tables cont.						
deg	sin(x)	cos(x)	tan(x)	cot(x)	csc(x)	sec(x)
18	0.3090	0.9511	0.3249	3.0777	3.2361	1.0515
19	0.3256	0.9455	0.3443	2.9042	3.0716	1.0576
20	0.3420	0.9397	0.3640	2.7475	2.9238	1.0642
21	0.3584	0.9336	0.3839	2.6051	2.7904	1.0711
22	0.3746	0.9272	0.4040	2.4751	2.6695	1.0785
23	0.3907	0.9205	0.4245	2.3559	2.5593	1.0864
24	0.4067	0.9135	0.4452	2.2460	2.4586	1.0946
25	0.4226	0.9063	0.4663	2.1445	2.3662	1.1034
26	0.4384	0.8988	0.4877	2.0503	2.2812	1.1126
27	0.4540	0.8910	0.5095	1.9626	2.2027	1.1223
28	0.4695	0.8829	0.5317	1.8807	2.1301	1.1326
29	0.4848	0.8746	0.5543	1.8040	2.0627	1.1434
30	0.5000	0.8660	0.5774	1.7321	2.0000	1.1547
31	0.5150	0.8572	0.6009	1.6643	1.9416	1.1666
32	0.5299	0.8480	0.6249	1.6003	1.8871	1.1792
33	0.5446	0.8387	0.6494	1.5339	1.8361	1.1924
34	0.5592	0.8290	0.6745	1.4826	1.7883	1.2062
35	0.5736	0.8192	0.7002	1.4281	1.7434	1.2208
36	0.5878	0.8090	0.7265	1.3764	1.7013	1.2361
37	0.6018	0.7986	0.7536	1.3270	1.6616	1.2521
38	0.6157	0.7880	0.7813	1.2799	1.6243	1.2690
39	0.6293	0.7771	0.8098	1.2349	1.5890	1.2868
40	0.6428	0.7660	0.8391	1.1918	1.5557	1.3054
41	0.6561	0.7547	0.8693	1.1504	1.5243	1.3250
42	0.6691	0.7431	0.9004	1.1106	1.4945	1.3456
43	0.6820	0.7314	0.9325	1.0724	1.4663	1.3673
44	0.6947	0.7193	0.9657	1.0355	1.4396	1.3902
45	0.7071	0.7071	1.0000	1.0000	1.4142	1.4142
46	0.7193	0.6947	1.0355	0.9657	1.3902	1.4396
47	0.7314	0.6820	1.0724	0.9325	1.3673	1.4663
48	0.7431	0.6691	1.1106	0.9004	1.3456	1.4945
49	0.7547	0.6561	1.1504	0.8693	1.3250	1.5243
50	0.7660	0.6428	1.1918	0.8391	1.3054	1.5557
51	0.7771	0.6293	1.2349	0.8098	1.2868	1.5890
52	0.7880	0.6157	1.2799	0.7813	1.2690	1.6243
53	0.7986	0.6018	1.3270	0.7536	1.2521	1.6616
54	0.8090	0.5878	1.3764	0.7265	1.2361	1.7013

Trigonometric Tables cont.						
deg	sin(x)	cos(x)	tan(x)	cot(x)	csc(x)	sec(x)
55	0.8192	0.5736	1.4281	0.7002	1.2208	1.7434
56	0.8290	0.5592	1.4826	0.6745	1.2062	1.7883
57	0.8387	0.5446	1.5399	0.6494	1.1924	1.8361
58	0.8480	0.5299	1.6003	0.6249	1.1792	1.8871
59	0.8572	0.5150	1.6643	0.6009	1.1666	1.9416
60	0.8660	0.5000	1.7321	0.5774	1.1547	2.0000
61	0.8746	0.4848	1.8040	0.5543	1.1434	2.0627
62	0.8829	0.4695	1.8807	0.5317	1.1326	2.1301
63	0.8910	0.4540	1.9626	0.5095	1.1223	2.2027
64	0.8988	0.4384	2.0503	0.4877	1.1126	2.2812
65	0.9063	0.4226	2.1445	0.4663	1.1034	2.3662
66	0.9135	0.4067	2.2460	0.4452	1.0946	2.4586
67	0.9205	0.3907	2.3559	0.4245	1.0864	2.5593
68	0.9272	0.3746	2.4751	0.4040	1.0785	2.6695
69	0.9336	0.3584	2.6051	0.3839	1.0711	2.7904
70	0.9397	0.3420	2.7475	0.3640	1.0642	2.9238
71	0.9455	0.3256	2.9042	0.3443	1.0576	3.0716
72	0.9511	0.3090	3.0777	0.3249	1.0515	3.2361
73	0.9563	0.2924	3.2709	0.3057	1.0457	3.4203
74	0.9613	0.2756	3.4874	0.2867	1.0403	3.6280
75	0.9659	0.2588	3.7321	0.2679	1.0353	3.8637
76	0.9703	0.2419	4.0108	0.2493	1.0306	4.1336
77	0.9744	0.2250	4.3315	0.2309	1.0263	4.4454
78	0.9781	0.2079	4.7046	0.2126	1.0223	4.8097
79	0.9816	0.1908	5.1446	0.1944	1.0187	5.2408
80	0.9848	0.1736	5.6713	0.1763	1.0154	5.7588
81	0.9877	0.1564	6.3138	0.1584	1.0125	6.3925
82	0.9903	0.1392	7.1154	0.1405	1.0098	7.1853
83	0.9925	0.1219	8.1443	0.1228	1.0075	8.2055
84	0.9945	0.1045	9.5144	0.1051	1.0055	9.5668
85	0.9962	0.0872	11.4301	0.0875	1.0038	11.4737
86	0.9976	0.0698	14.3007	0.0699	1.0024	14.3356
87	0.9986	0.0523	19.0811	0.0524	1.0014	19.1073
88	0.9994	0.0349	28.6363	0.0349	1.0006	28.6537
89	0.9998	0.0175	57.2900	0.0175	1.0002	57.2987
90	1.0000	0.0000		0.0000	1.0000	

TUBE STAGGERING

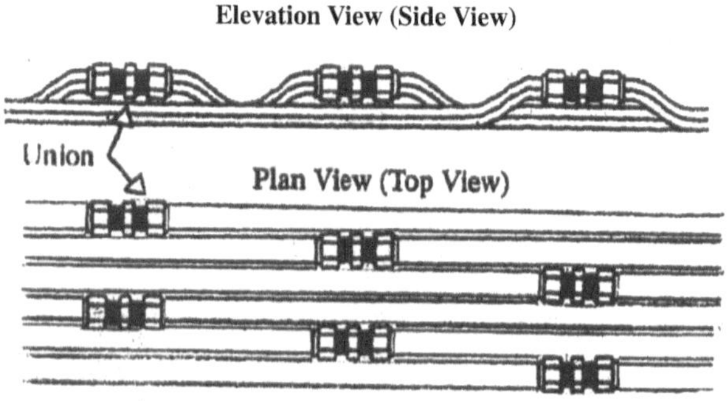

Figure 12 - Tube Stagger

These tubes with 45 degree bends are shown from the side and then from above. The 45 degree bends are bent up and then back parallel with the tube run after which a tube union will be installed, and then two more back to back 45 degree bends will be bent, so that the tube will proceed onward in the tube run.

Due to space limitations, tube unions cannot be installed side by side. Offsets need to be staggered to accommodate the tube unions. A 3 tube repeating stagger is the best stagger layout design to accommodate the bulkiness of the tube unions, to make the tube unions accessible for the use of wrenches, and to make the tube stagger look organized and neat. Every 4th tube will start out at the same relative point in the overall tube layout as the first tube. This will be the start of the next 3 tube stagger repeat pattern of bends.

TUBE STAGGER MEASUREMENT

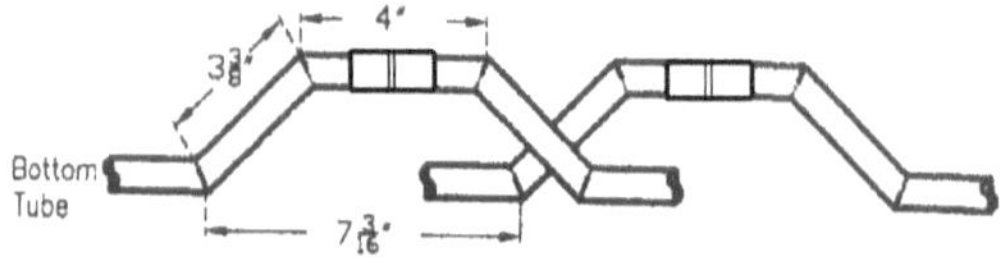

Figure 13 - Tube Stagger Measurement

To arrive at the tube stagger measurement is simple. The tube to calculate the tube stagger from is the first tube in the lower run. The measurement over the bottom tube union in the tray center to center between the two top 45 degree bends is 4". Referring to figure 14, the total spread between the 3 tube unions is 2 – ¼" (in the area of the tube stack, 1 – ⅛" + 1 – ⅛" = 2 – ¼"). Since the bends are 45 degree bends, multiply the spread by 1.414 to arrive at the hypotenuse (travel) of the spread, which equals 3.1815 (the nearest 1/16" is 3 – 3/16"). Add this number 3-3/16" to 4", (the measurement over the bottom union in the tray center to center between the two top 45 degree bends). Each tube will now start to bend at 7 – 3/16" further along the tray than the preceding tube. This will be the tube stagger for all tubes in all layers. All tubes will stagger up out of the tray and back down into the tray uniformly. This entire layout is extremely orderly and eye appealing.

TUBE STACKING

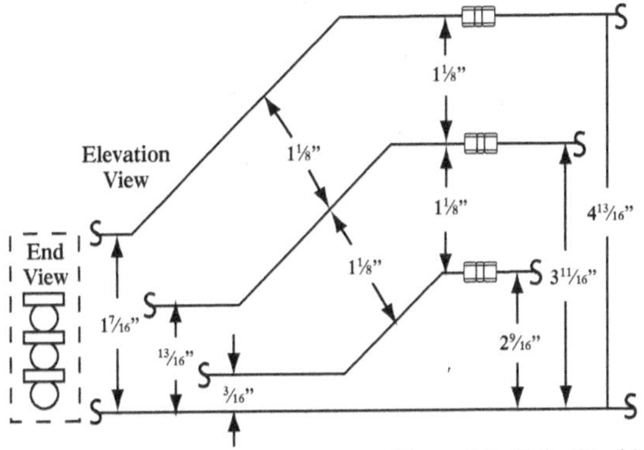

Figure 14 - Tube Stacking

Tube stacking is one of the more difficult aspects of tube layout to describe. The following explanation along with the preceding diagram will alleviate any apprehension of being capable of laying out a neat, orderly tube stack.

To begin, sketch the 3 tubes stacked one atop another with a ¼” space between the tubes to simulate the hold down bar which would be installed above each layer of tube. The bottom most 3/8” O.D. tube runs will be at 3/16” on center (O.C.) above the inside base of the tray, the middle tube runs will be at 13/16” O.C. above the inside base of the tray, and the top tube runs will be at 1 – 7/16" O.C. above the inside base of the tray. These are the 3 elevations above the inside base of the tray that the 3 straight tube layer runs will travel.

Now, lay the 3 tube unions next to one another on a table and with the wrenches used to tighten the tube unions, place the 3 tube unions slightly apart until the tightening wrench can move enough to tighten the nut onto the tube union body of the middle tube union; this is the tube union which will be the most difficult to access with tube wrenches. In this case, with an 11/16" open end wrench, the spread between the tubes will measure 1 – 1/8". Use this measurement to keep a spread between the centerline of the top tube of the 3 straight tube run layers in the tray and the bottom tube union. This spread will also be used as a center to center measurement between the tube unions. From the inside bottom of the tray to the centerline of the three tube unions will measure 2 – 9/16" for the bottom tube union, 3 – 11/16" for the center tube union, and 4 – 13/16" for the top tube union. (See Figure 15)

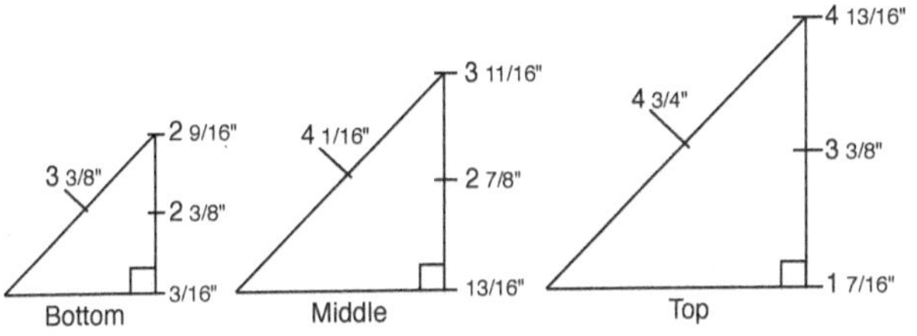

Figure 15 - 3 Tube Travel Lengths

Figure 15 shows the measurements in each of the three tube stack layers. The following is an explanation of the offset for the bottom tube. Centerline of the bottom tube union above the base of the tray is 2 – 9/16". Centerline of the bottom tube bolted securely into the tray is 3/16". When we subtract 3/16" from 2 – 9/16" the remainder is 2 – 3/8". Since we are bending 45 degree bends, we know that both the set and the run are 2 – 3/8". The length of the travel of any 45 degree right triangle is always 1.414 times the length of either the set or the run. When 2 – 3/8" is multiplied by 1.414, we find that the travel for the lower tube between the two 45 degree bends equals 3 – 3/8". The set, run, and travel for both the middle and top offsets would be calculated in the same manner.

TUBE JOINING

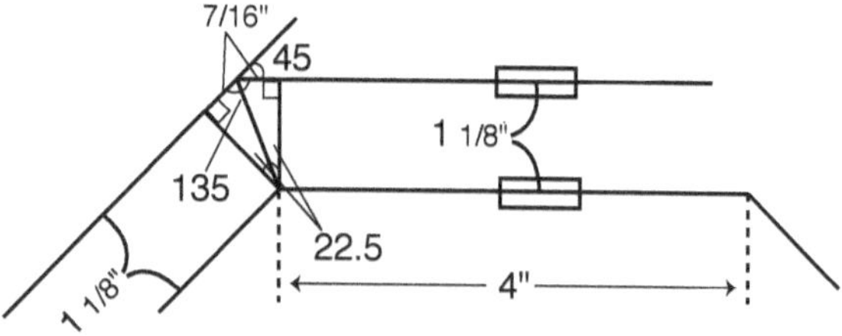

Figure 16 - Tube Joining Calculation

The tube tail on each side of the bottom tube union is 1 – 7/8" plus 1/8" to the center of the tube union (Figure 10, ½ of ¼" = 1/8"). This equals 2" from the center of the bottom tube union to the center of either 45 degree bend. The dimension of 4" is determined to be an adequate length for the bottom tube from center to center of the 45 degree bends over the bottom tube union. Next the length of the middle and top tubes of the 3 tube joining areas must be calculated.

By referring to Figure 14, we know that the spread between the three stack layers is 1 – 1/8" equal spread at the top and the adjoining travel, and the bends are 45 degree bends; the remainder of the bend on one side of a straight line is 135 degrees (Figure 17). When we bisect the 135 degree bend, we end up with 67 – ½ degrees at that corner of a right triangle. If we subtract this 67 – ½ degrees from 90 degrees, we end up with 22 – ½ degrees at the bottom corner of the right triangle. We now know the angle and the run, and can calculate for the length of the set. Using the tangent of angle formula, (formulas after Figure 11 page 44) the length of the set will be calculated and the length of the middle and top tube runs center to center between the 45 degree top bends over their respective tube unions will be determined.

Example:

Set = run x the tangent of the angle.

Set = run x the tangent of 22 – ½ degrees.

Set = (1 – 1/8") x 0.414

Set = 1.125 x 0.414

Set = 0.4659 (Table 1 to the nearest 1/16")

Set = 7/16"

Because the spread between the tube unions is 1 – 1/8" (Figure 16), the middle tube union length is 4" + 7/16" +

7/16" to equal 4 – 7/8" center to center between the two top 45 degree bends. The top tube union length is (4 – 7/8") + 7/16" + 7/16" to equal 5 – ¾" center to center between the two top 45 degree bends.

TUBE STACK SETBACK

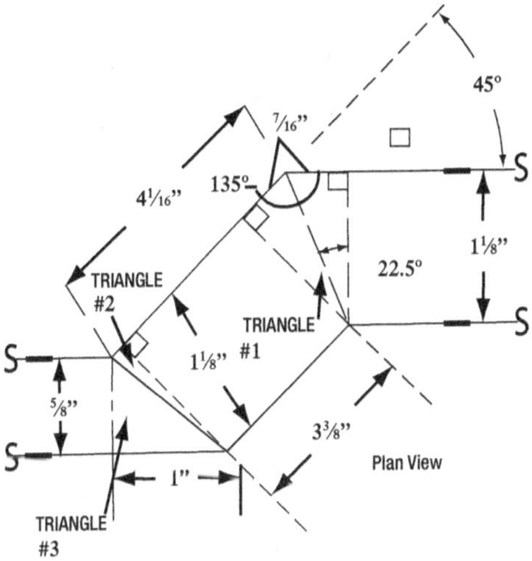

Figure 17 - Tube Stack Setback

The length of the shortest side in triangle #2 is easily calculated. The lower tube travel length center to center is 3 – 3/8". The middle tube travel length center to center is 4 – 1/16". The set on triangle #1 = 7/16". When we add 7/16" to 3 – 3/8", we get 3 – 13/16". We will then subtract 3 – 13/16" from 4 – 1/16" to find a difference of ¼". This is the length of the shortest side of triangle #2. We also know the total angle of turn of triangles #2 and #3 is 90 degrees plus 45 degrees to equal 135 degrees of turn. We will now calculate the angle of triangle #2 at the bend to determine the set of triangle #3, which will be the starting point set back of each layer of tubes above one another.

Since we know the length of both the set and the run (triangle #2), we will use the tangent of angle formula (formulas are located after Figure 11 page 44).

Example:

The tangent of the angle = set/run

The tangent of the angle = (1 – 1/8) divided by ¼

The tangent of the angle = 1.125/0.25

The tangent of the angle = 4.5 (tangents from the Trigonometry Tables, this angle is 77.5 degrees)

Since triangle #2 is 77.5 degrees at the corner next to the ¼" run, when 77.5 degrees is subtracted from the total turn of 135 degrees, the adjoining angle (in triangle #3) is 57.5 degrees. The run on triangle #3 is the same as the spread between the two tubes which is 5/8". Once again using the tangent of angle formula:

Set = run x the tangent of the angle

Set = 5/8" x the tangent of 57.5 degree (tangents from the Trigonometric Tables)

Set = 0.625 x 1.57

Set = 0.98125 (Table 1 – the nearest 1/16")

Set = 1"

TUBE STACK LAYOUT

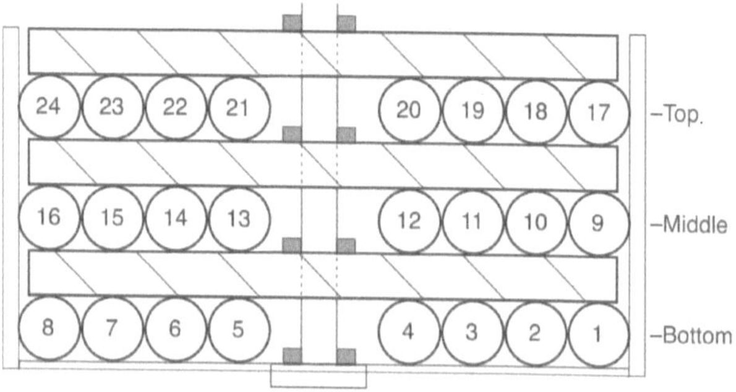

Figure 18 - Tube Stack Layout

Above is an end view of a 4" tray, illustrating 8 each 3/8" O.D. tubes per layer, with a ¼" x 1" x 3 – ¾" aluminum flat bar above each layer to solidly bolt the layer into place before installing the next layer of tubes. The ¼" bolt going through the exact center of the aluminum flat bar is at least 2 -1/2" long and has a nut securing the bolt at each layer, and on the top. There will be 4 nuts total per bolt and the bolts should be carriage bolts.

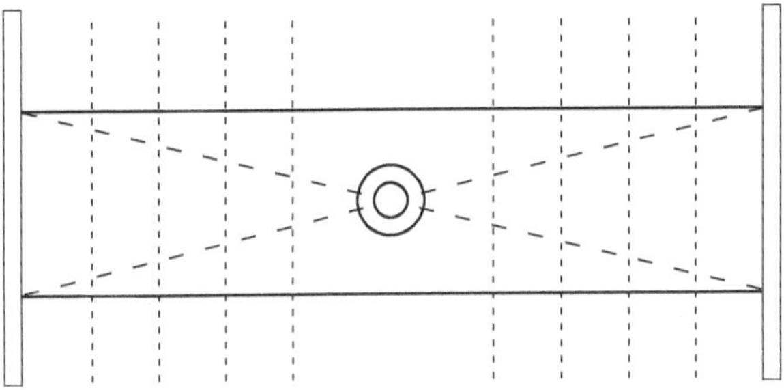

Figure 19 - Flat Bar Top View

In figure 19, the 4" tray is being viewed from the top and what is being seen is the 1" wide by 3 -3/4" long aluminum flat bar with a ¼" hole in the exact center for the ¼" carriage bolt.

COMPOUND BENDS

Anytime 1 bend is bent in such a manner as to do what 2 bends would usually do, a compound bend is required. The following are two separate examples of how to calculate compound bends:

Example #1: Many times a tube will be pitched up an incline and then will come to a 90 degree corner (viewing the tube layout in the elevation view), after which the tube will continue pitching up at an incline. The bend at the intersecting corner will not be a 90 degree bend!

How do you figure out the degree of this bend?

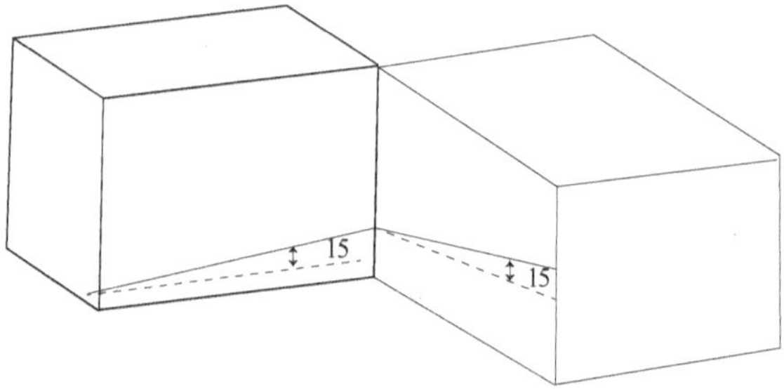

Figure 20 – Compound Bend – Adjoining Planes

A tube slopes up an incline at 15 degrees where it comes to a 90 degree corner (viewing the tube layout in an elevation view) and then continues sloping up at 15 degrees. Use the following formula:

(the sine of the angle – slope #1) x (the sine of the angle – slope #2) = the cosine of the angle of the bend

(the sine of 15 degrees) x (the sine of 15 degrees) = the cosine of the angle of the bend

(.2588) x .2588) = .06697744 (86 degrees)

This formula will work with any degree of slope (from 1 to 89 degrees) in slope #1 multiplied by any degree of slope (from 1 to 89 degrees) in slope #2. If either adjoining plane has no slope, then the bend where the 2 tubes intersect will be a 90 degree bend.

By slightly changing the preceding formula, the angle of the bend can be calculated where 2 tubes in 2 perpendicular planes meet at a given point!

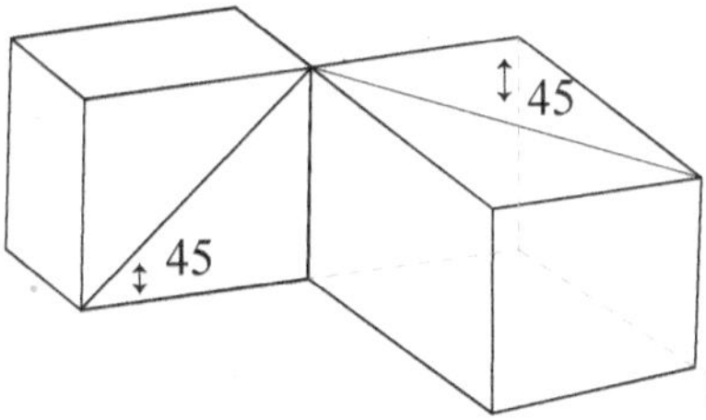

Figure 21 – Compound Bend – Perpendicular Planes

Example #2: A tube pitches up an incline at 45 degrees (plane #1), at some point the tube bends at a 45 degree angle into a horizontal plane (plane #2). The bend where the two tubes intersect will not be a 45 degree bend.

How do you figure out the degree of this bend?

(the cosine of the angle – plane #1) x (the cosine of the angle – plane #2) = the cosine of the angle of the bend

(the cosine of 45 degrees) x (the cosine of 45 degrees) = the cosine of the angle of the bend

(.7071) x (.7071) = .4999 (60 degrees of bend)

This formula will work with any degree of pitch (from 1 to 89 degrees) in the vertical plane (plane #1), multiplied by any degree of roll (from 1 to 89 degrees) in the horizontal plane (plane #2).

COMPOUND LENGTH AND BEND

The following example should be used as a template to calculate any 3 dimensional compound bend which will be needed with any degree of rise/drop (from 1 to 89 degrees) along with any degree of bend right/left (from 1 to 89 degrees).

The easiest formula to use to find the diagonal length (D) through any 3 dimensional rectangular box is:

$$D = \sqrt{A^2 + B^2 + C^2}$$

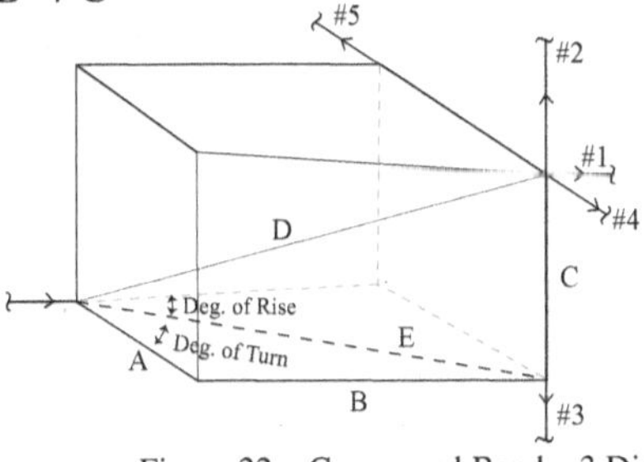

Figure 22 – Compound Bend – 3 Dimensional

Example: A = 40
B = 35
C = 17

$$D = \sqrt{(40)^2 + (35)^2 + (17)^2}$$
$$D = \sqrt{1600 + 1225 + 289}$$
$$D = \sqrt{3114}$$
$$D = 55.80$$

When calculating the degree of bend of this compound bend, since the tube is traveling through an imaginary rectangular box, three dimensions will be taken into consideration!

Dimension #1, how much the bend is rising or falling (degree of rise).

Dimension #2, how much the bend is bending left or right (degree of turn), and;

Dimension 3, how much the bend is advancing.

To calculate the degree of this compound bend:

First, to factor in dimension 3 for how much the tube is advancing: the travel (E) of the right triangle with sides A and B must be calculated by using the following formula.

$$E = \sqrt{A^2 + B^2}$$
$$E = \sqrt{(40)^2 + (35)^2}$$
$$E = \sqrt{1600 + 1225}$$
$$E = \sqrt{2825}$$
$$E = 53.15$$

Second, calculate the degree of both angles (rise/fall and turn) by using the formula:

Cosine of the angle = Run / Travel
Cos. of the rise/fall is 53.15 / 55.80 = .9525 (17.73 degrees)
Cos. of the turn is 40 / 53.15 = .7526 (41.18 degrees)

The formula needed to calculate the degree of this compound bend is (cos. x cos. = cos.). The cosine of the turn times the cosine of the rise/fall will equal the cosine of the compound bend.

Compound bend is .9525 x .7526 = .7168 (Cos. of 44.21 degree)

The compound bend in this example will be a 44.21 degree bend traveling through three planes; forward, up and to the right.

There are 5 common directions in which the tube could proceed upon exiting the area of the imaginary rectangular box which was illustrated for the previous rolling offset.

#1 – Straight forward. As with any back to back offset, if after the offset, the run of the tube keeps advancing in the direction parallel with the original tube run, the second bend will be the same as the bend previously calculated (44.21 degrees).

#2 – Straight up. This bend will be the complement (an angle related to another so that the sum of their measures is 90 degrees) of the degree of the rise of the bottom elbow. The rise was 17.73 degrees, so 90 degrees minus 17.73 degrees = 72.27 degrees.

#3 – Straight down. This bend will be the same as the degree of the rise plus 90 degrees (1/4 of a circle). 17.73 degrees plus 90 degrees = 107.73 degrees.

Since #2 and #3 proceed along a straight line in opposite directions from one another, that line will represent the diameter of a circle, and the 2 previous numbers will add up to 180 (½ of a circle).

#4 – Bend to the right. This bend will be the complement of the degree of the turn of the bottom elbow which was 41.18 degrees, times the rise.
90 degrees minus 41.18 degrees = 48.82 degrees of turn.
The degree of the turn will be calculated by using the formula:
cos. x cos. = cos.
(cos. of 48.82 degrees = .6584) x (cos.of 17.73 degrees = .9525) = .6271 (cos. of 51.2 degrees)

#5 – Bend to the left. This bend will be the same as the bottom degree of turn, plus 90 degrees (¼ of a circle),

times the rise.
90 degrees plus 41.18 degrees = 131.18 degrees of turn.
The degree of turn will be calculated by using the formula:
cos. x cos = cos.
(cos. of 131.18 degrees [-.6584]) x (cos. of 17.73 degrees [.9525]) = -.6266 (cos. of 128.8 degrees)

Since #4 and #5 proceed along a straight line in opposite directions from one another, that line will represent the diameter of a circle, and the 2 previous numbers will add up to 180 (½ of a circle).

Note: In this chapter and the previous chapter, a calculator was used for all math equations and to find the trigonometric degrees.

VIBRATION LOOP

In many commercial applications, for precise instrument control, engineers are detailing the installation of a 360 degree vibration loop in each tube run immediately before where the tube connects to the controlled instrument.

By using this bending method, a 360 degree elliptical vibration loop can quickly and easily be laid-out and bent in any tube run while using the tube bender. **Do not bend the tube around a pipe by hand.** Anyone who makes any bend around a pipe by hand does not care enough to do work for themselves or anyone else on any job. The proper bending tool has been invented and all experienced tube bending mechanics will use the proper tool for all bending jobs so that all bends will be consistent, identical, and eye appealing. An elliptical shape for the vibration loop is bent by inserting a short travel piece between the two 180 degree bends, so that the tube bender can easily be removed and repositioned after each 180 degree bend.

The following example will only take 5 minutes to layout, mark, and bend; and every vibration loop will be identical to every other vibration loop which is bent.

Given: ½" O.D. tube being bent with a 1½" radius bender.

Bend:

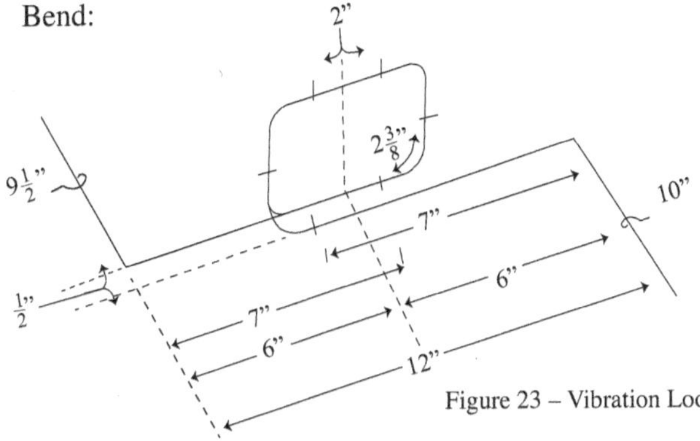

Figure 23 – Vibration Loop

The tube which is needed for the preceding illustration is: 9½" + 9⅜" + 6¾" + 9⅜" + 10" – (two each 90 degree tube gains).

In the preceding example; tube gain for the two 180 degree bends does not need to be considered in the equation, because the arc length measurement (2⅜") of each 90 degree quadrant is used.

Tube Schedule
1 – 9½" – E-C of a 90 degree bend
2 – 9⅜" – C-C of a 180 degree bend
3 – 6¾" – C-C of a 180 degree bend
4 – 9⅜" – C-C of a 90 degree bend
5 – 10" – C-E of tube

The total length tube needed = (45") – 1¼"
Total length tube needed = 43¾"

To easily center the vibration loop between each leg, dimension number 2, the first 9⅜" dimension is arrived at by adding 3 measurements together.
6" (½ of the 12" dimension between the two 90 degree legs)
1" (½ of the 2" travel piece between each 180 degree ½ circle)
2⅜" (the arc length of one 90 degree bend [½ of a 180 degree bend])

Dimension number 3 is the arc length of 2 each 90 degree bends plus the 2" measurement between each 180 degree bend to be able to complete bending the elliptical vibration loop.

While sketching a vibration loop, the travel length of the tube from the start of the bend to the center of the first 180 degree bend will always be whatever the A.C. measurement is for a 90 degree bend.

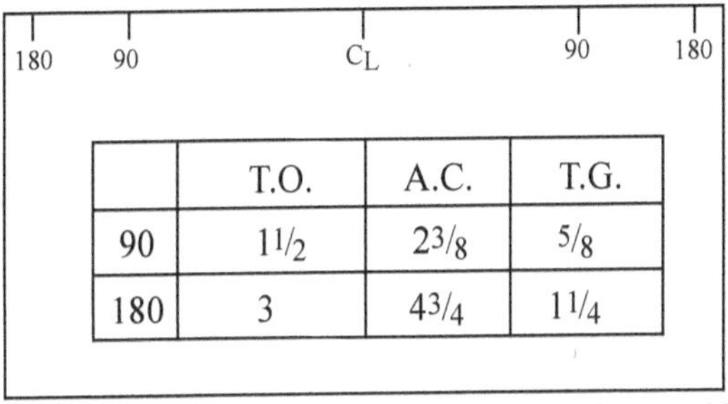

	T.O.	A.C.	T.G.
90	1 1/2	2 3/8	5/8
180	3	4 3/4	1 1/4

Figure 24 – Tubespoon Marker
Not to Scale

To bend this vibration loop, layout and mark the entire 43¾" long tube. While bending the vibration loop; since the tube will need to be rotated through a complete 360 degree plane, the vibration loop should be bent before any other bends are bent for the layout on that tube.

BEND TEST

Given 46" of ½" O.D. tube using a 1½" radius bender

Bend the following:

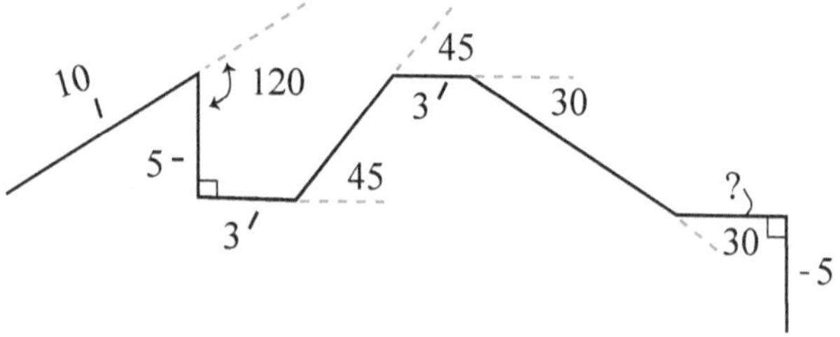

Figure 25 – Bend Test

To solve: <u>1st</u>, construct a tubespoon marker for all bends. For a 120 degree bend; the T.O., A.C., and T.G. measurements are calculated by adding a 90 degree bend and a 30 degree bend together.

120 90 45 30 C_L 30 45 90 120

	T.O.	A.C.	T.G.
90	1 1/2	2 3/8	5/8
45	5/8	1 3/16	1/16
30	3/8	3/4	0
120	1 7/8	3 1/8	5/8

Figure 26 – Tubespoon Marker
Not to Scale

2nd, mathematically calculate the length of travel dimension #4 with back to back 45 degree bends and a set of 5", and travel dimension #6 with back to back 30 degree bends and a set of 5".

Turn to the formulas after Figure 11 page 44. When one angle and 1 side (the set) are known; to calculate the length of the travel use the formula, travel = set x the cosecant of the angle.

Travel #4 = 5" x 1.414 (Csc. of 45 degrees)
Travel #4 = 7 1/16"
Travel #6 = 5" x 2.000 (Csc. of 30 degrees)
Travel #6 = 10"

3rd, use the information from the tubespoon marker to calculate the total tube gain for this bend test.

2 each 90 degree bends at $\frac{5}{8}$" per bend	= $1\frac{1}{4}$"
2 each 45 degree bends at $\frac{1}{16}$" per bend	= $\frac{1}{8}$"
2 each 30 degree bends at 0" per bend	= 0"
1 each 120 degree bend at $\frac{5}{8}$" per bend	= $\frac{5}{8}$"
Total tube gain for bends in test (Figure 25)	= 2"

4th, to calculate the length of the (?) dimension; add the other 7 measurements together.

10" + 5" + 3" + $7\frac{1}{16}$" +3" + 10" + 5" = $43\frac{1}{16}$"
then subtract that total from 46"
46" – $43\frac{1}{16}$" = $2\frac{15}{16}$"

Finally add that reminder ($2\frac{15}{16}$"), to the tube gain to get the length of (?)
(?) = $2\frac{15}{16}$" + 2"
(?) = $4\frac{15}{16}$"

5th, construct a tube layout schedule

#1 E-C = 10" at 120 degree
#2 C-C = 5" at 90 degrees
#3 C-C = 3" at 45 degrees (right-hand bend)
#4 C-C = $7\frac{1}{16}$" at 45 degrees
#5 C-C = 3" at 30 degrees

#6 C-C = 10" at 30 degrees
#7 C-C = 4 15/16" at 90 degrees
#8 C-E = 5"

6th, make a bending template.

Lastly, bend the test diagram. Anyone who successfully bends this test will easily pass any bending test!

SUMMARY

1. The radius of the bender which is being used dictates all T.O. and A.C. measurements.
2. T.O. = the tangent of ½ of the degrees of the bend x the radius of the bender (Up to 90 degrees). Above 90 degrees the T.O., A.C. and T.G. are accumulative starting with the measurements for (1) 90 degree bend.
3. A.C. = the degrees of the bend x 0.01745 x the bender radius.
4. Formulas for all bends will be located after Figure 11 page 44.
5. Measure and mark accurately.
6. Double check all bends for proper degree and all measurements for proper length.
7. Use a bending template to keep track of changes in plane.
8. Align the tube properly in the bender.
9. All bends will be bent with the "0" mark of the bending handle on one of the outside lines on the tube at each bend.
10. Take pride in your work, make every job look like it was done by the most knowledgeable and the most qualified person.

CONCLUSION

This book is the sole reference needed by anyone who wishes to become an accomplished tube bender. The information contained herein will relate to any size conduit, pipe, or tube in any layout. Once these principles are committed to memory, not only will you have become a great tube bender, you will know all of the basics of a great pipefitter layout mechanic!

Good Luck and Great Bending

NOTES

NOTES

www.ingramcontent.com/pod-product-compliance
Lightning Source LLC
LaVergne TN
LVHW101617210926
848480LV00009B/715
9798894199580